ཀྱངས་ཀའི་སྒྲུང་།

THE NUMBER STORY

SMALL BOOK ONE

ENGLISH - TIBETAN

Numbers Teach Children Their Number Names

written and illustrated by

MISS ANNA

Early Reader Edition of *The Number Story 1*
Bronze Medal Winner, 2016 Wishing Shelf Book Award

LUMPY Publishing

Library of Congress Control Number: 2018902040

Names: Miss Anna, author.
Title: Number story : numbers teach children their number names / Miss Anna.
Description: Portland, OR: Lumpy Publishing, 2018.
Identifiers: ISBN 978-1-949320-16-9 | LCCN 2018902040
Summary: The pictures and rhymes present stories which introduce numbers 0-10.
Subjects: LCSH Numeration—English—Tibetan--Pictorial works--Juvenile literature. | BISAC JUVENILE NONFICTION /
Languages: English—Tibetan
Classification: LCC QA141.3 .M57 2018 | DDC 513—dc23

Publisher: Lumpy Publishing
Website: www.missannabooks.com
Email: missanna@missannabooks.com

Paperback: ISBN 978-1-949320-16-9
Printed in the U.S.A.   1 3 5 7 9 10 8 6 4 2

Want to learn our number names?

ཁྱེད་རང་གྲངས་ཀའི་མིང་རྣམ་སྦྱང་འདོད་ཡོད་པས།

It is very easy and a lot of fun!

དེ་ནི་ཤིན་ཏུ་ལས་སླ་པོ་དང་དགོད་བྲོ་ལྡན་པ་ཞིག་ཡིན།

Say-along our little jingle

ང་ཚོ་ཚང་མ་ལྷན་དུ་སྐྱུང་དེ་གློག་གི་ཡིན།

starting from Number One!

ང་ཚོ་གྲངས་ཀ་དང་པོ་ནས་རྩུ་འཆུག་གི་ཡིན།

1

ONE looks like my one finger.

དཔེར་ན་ངའི་མཛུབ་མོ་གཅིག་ལྟ་བུ།

1
ONE!
বাউৱা

2

TWO   trails a tail.

༢ ☆ གཉིས་

ང་མའི་རྗེས་ཤུལ།

A TAIL! ཧ་མ།

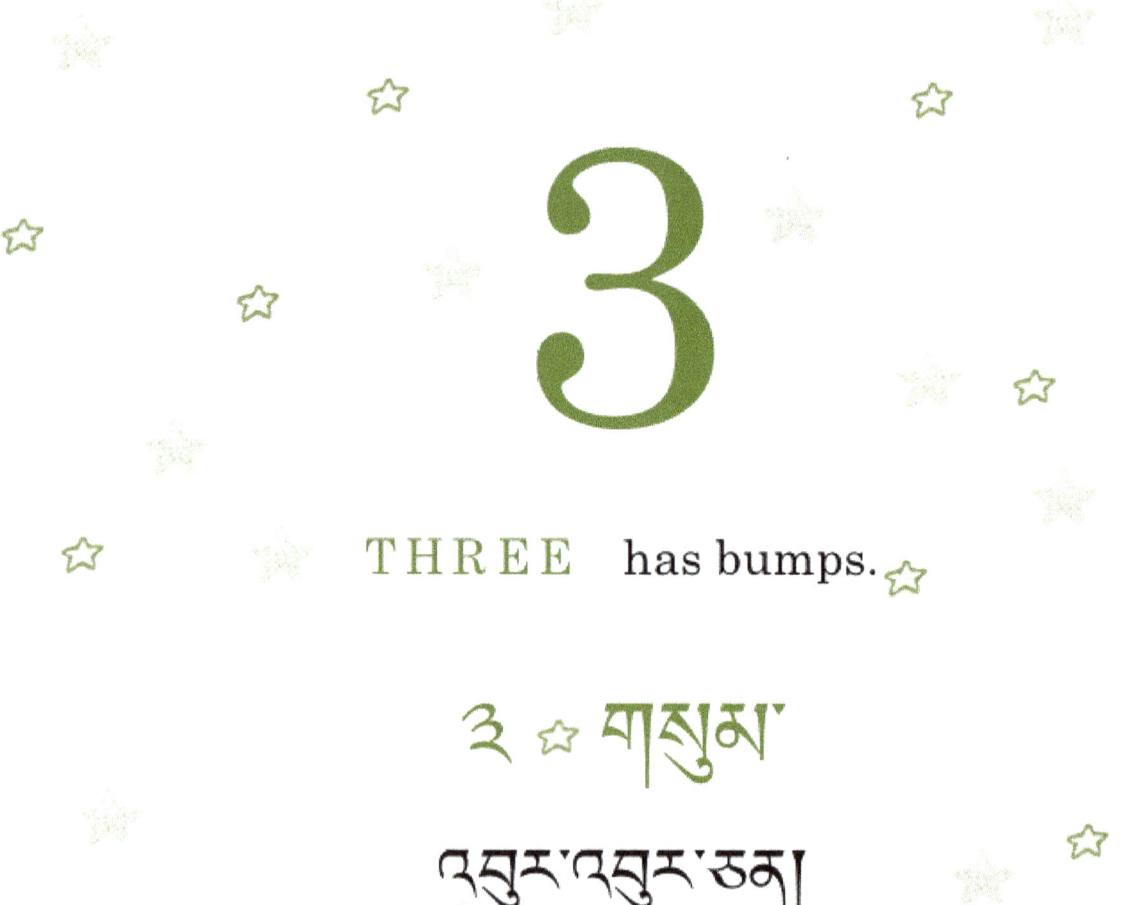

3

THREE has bumps.

༣ ☆ གསུམ་

འབུར་འབུར་ཅན།

BUMPY! འབར་རེ་འབུར་རེ་ལྷུན་པ།

4

FOUR carries a sail.

གཞི།

གྲུའི་རྐྱང་གཡོར་འཁྱུར་བ་ལྔ་བ།

4
A SAIL!
གྲུའི་རྐྱང་གཡོབ།

5

FIVE is a racing track.

༥ ལྔ་

རྒྱུག་ལམ་ལྟ་བུའི་དབྱིབ་ཅན།

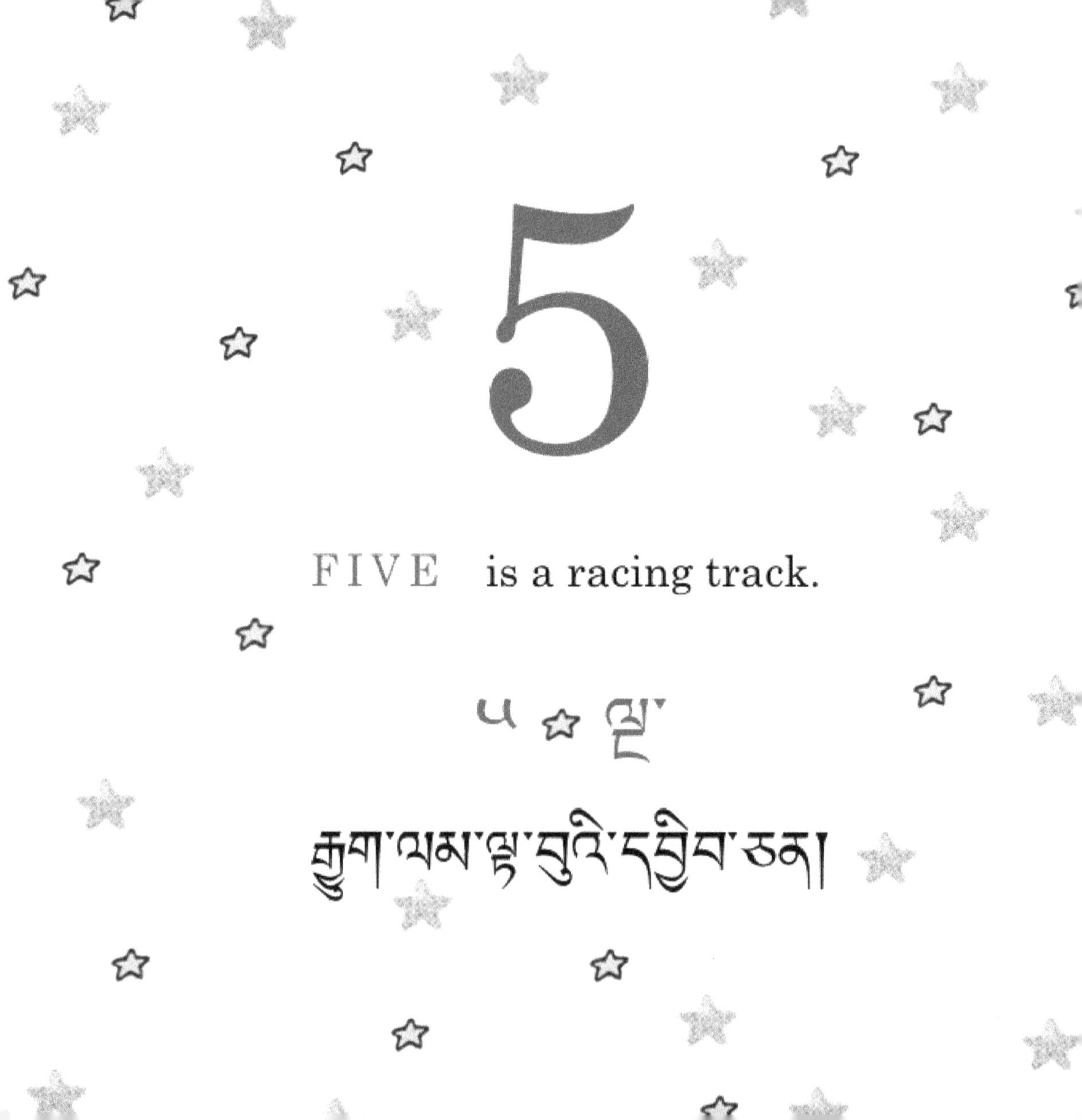

VROOM
VROOM

6

SIX curves like a snail.

༦ ☆ དྲུག

གུག་ཕྱིག་ལྟར་བ། དཔེར་ན་སྦྲུག་པའི་བུ་མོ་ལྟ་བུ།

A SNAIL! སྦུལ་པའི་བུ་མོ།

7

SEVEN   has a sharp angle.

བདུན་

ཟུར་ཁུག་ཅན་ལ་ཆེ་རྐོ་བ།

BE CAREFUL! IT'S SHARP!
གཟབ་གཟབ་གནང་རོགས།

8

EIGHT is rollercoaster rails.

ཅེ་ལ་ན་ལགས།
YIPPEE!

9

 is a bubble on a stick.

ཁ་དགུ་

དགུག་པའི་ཅེ་ལ་ཆོག་ཆོག་ཡོད་པ་ལྟ་བུ།

ཆུ་བག་ཆུ་བག་ཚ་ཀ།

10

༡༠ ☆ བཅུ་

ཉ་ཆེན་གྱི་མིག་གཅིག་གི་དབྱིབས་ལྟ་བུ།

HELLO!
བཀྲ་ཤིས་བདེ་ལེགས།།

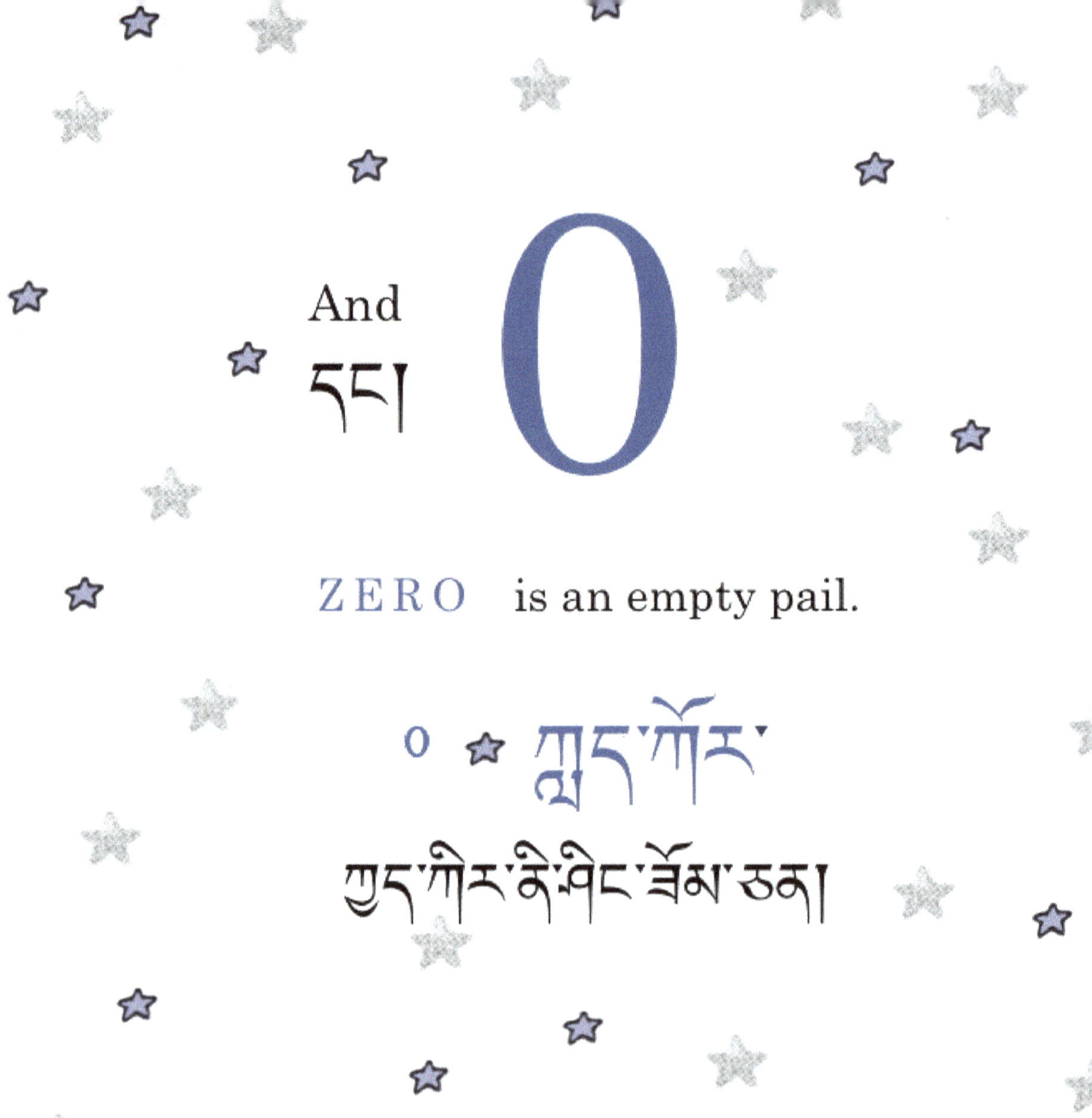

And

དང་།

0

ZERO   is an empty pail.

0

ཀླད་ཀོར་

ཀླད་ཀོར་ནི་ཁེང་ཆོམ་ཅག

IT'S
EMPTY!
སྟོང་པ་ཚ།

Thank you for playing with us today.

We had a lot of fun too!

ང་ཚོ་ལྷན་དུ་ཉེད་གྲོགས་གནང་བ་ལ་ཐུགས་རྗེ་ཆེ་ཞུ།

ང་ཚོ་སྐྱོ་བ་ཆེན་པོ་བྱུང་།

We are your Number friends,
Zero to Ten,
Who will be here for you~

ང་ཚོ་ཁྱེད་ཀྱི་གྲངས་ཀའི་རོགས་པ་ཆག་པ་རེད།
ང་ཚོ་ཁྱེད་ཀྱི་གྲངས་ཀ་གཅིག་ནས་ཀླད་ཀོར་བར་གི་གྲོགས་འགྱུར་བ་རེད།
རྒྱུན་ནས་དུ་ང་ཚོ་ཁྱེད་ཀྱི་ཅེད་དུ་གནས་ཡོད།

Bye-bye now!
See you again soon!

ལེགས་པོ་ཨོ་ནས་ཕེབ་རོགས་གནང་།
མྱུར་དུ་འཇལ་བའི་སྐོན་ལམ་ཡོད།

The Numbers are *SINGING* too!

To sing-a-long, look for Miss Anna Number Story
at your favorite music store like iTUNES.

MP3

Numbers 0-10
IDENTIFYING
& COUNTING

Numbers 11-20
& Ordinals

first, second, third...

Numbers 0-100
& Place Values

ones, tens, hundreds...

About Clocks
& Telling Time

hours, minutes, seconds

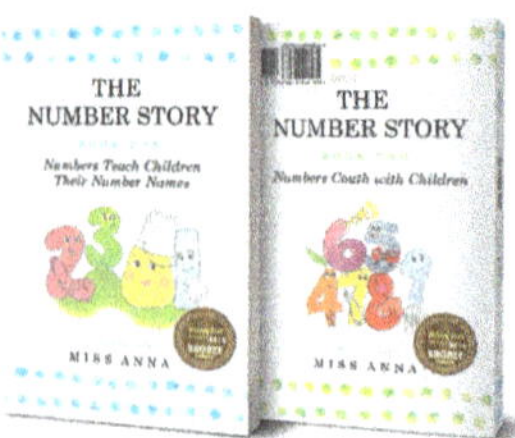

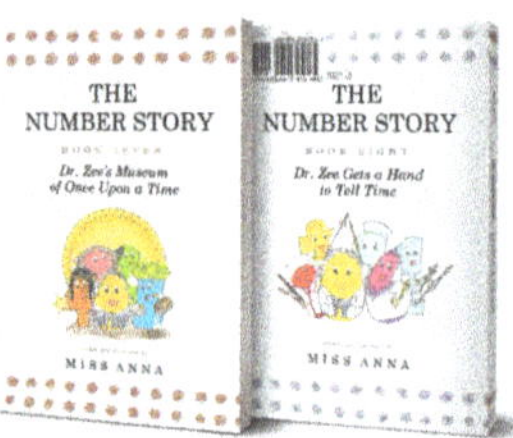

Number Story 1 & 2

isbn: 978-0-996216-48-7

Number Story 3 & 4

isbn: 978-1-945977-01-5

Number Story 5 & 6

isbn: 978-1-945977-06-0

Number Story 7 & 8

isbn: 978-1-949320-40-4

For more Miss Anna books to love,
visit us at

w w w . m i s s a n n a b o o k s . c o m

Numbers are working hard all over the world!
Come Travel the World with Us!

www.ingramcontent.com/pod-product-compliance
Lightning Source LLC
Chambersburg PA
CBHW040902070726
47599CB00035B/2270